WEATHERLY

ANIMAL NATURE

THE ART OF JOE WEATHERLY

VOLUME 2

ALSO BY JOE WEATHERLY

JOE WEATHERLY SKETCHBOOK VOL 1
ANIMAL ESSENCE THE ART OF JOE WEATHERLY VOL 1
THE WEATHERLY GUIDE TO DRAWING ANIMALS

 Many of the images in this book were for an exposition at the Galerie Daniel Maghen, Paris France

Published by
Weatherly Studio
263 North Bush Street
Santa Ana, CA 92701

Book layout and design by Weatherly Studio

ISBN 978-0-9710314-4-9 (Hardcover_First Edition)

Edited by Laura West

Printed in China

First Printing

www.joeweatherly.com

ACKNOWLEDGEMENTS

I would like to thank my supportive and talented wife Jimin for all of her encouragement and help with designing this book. Her trusty and informed eye has also gotten me out of many problem areas with composition and color in my paintings. Whenever she gives me advice to change something, it always looks better. She supports my dreams, artistic ambitions, and the passion I have for the animal kingdom. A painting lifestyle can involve pursuing one's dreams with lonely dedication; luckily for me, I have someone to go with me on the journey. I am truly fortunate.

A big thanks goes out to my mother for supporting my craft early on, and continuing to do so today. Parents who support their children's dreams are very special indeed.

To the very talented David Colman, who started off as a student of mine many years ago, and went on to become one of my best friends: thanks for all of the great times past, present and future. I couldn't think of a better person to journey with through the art of the animal kingdom.

Also, thanks to Steve Hampton for many years of friendship and support. This amazing draftsman and painter gave me many helpful bits of information on the art of picture-making, color, and composition.

To Mark, Fabian, Chris, and Nicholas, for making the cons extra special.

To Johnny and Josh, for those special art conversations over a pint of Abbot Ale.

A big thanks goes out to renowned archaeologist and Mayan expert Peter D. Harrison for his love of the jaguar, and hence, his collection of my biggest paintings. Peter is an example of a great collector who acquires art for the love of the subject matter. I am fortunate that tropical animals engaged in action are as fascinating to him as they are to me.

To Jane McMillan, for all of the years of friendship, trips overseas, and for being one of my best collectors.

Cheers to Nick and Arno from Pulp's in Paris for bringing my books into Europe and expanding my market.

Special gratitude goes out to Daniel, Olivier, Anabelle, Benoit, and Didier at the Galerie Daniel Maghen in Paris, France, for giving me the opportunity to show my work in Europe. Much of the work made for that exposition fills the pages of this book. It was an amazing experience to work with the gallery, as well as with such professional and simply amazing people.

Lastly, I'd like to thank the many zoo animals who, whether they knew it or not, were my most precious models.

For the Galerie Daniel Maghen

Joe Weatherly is a Southern California based artist specializing in the drawing and painting of animals. His style is bold and vigorous capturing the essence and drama of the subjects he draws and paints. The attitude and expression of the animal's character along with telling a visual story is what his work conveys. Conservation of the natural world is something Joe is very passionate about and hopes his work will motivate people to protect it and promote its survival. Books that Joe Weatherly has published are the Weatherly Guide to Drawing Animals, Animal Essence the Art of Joe Weatherly, and the Joe Weatherly Sketchbook. Teaching drawing is a part time endeavor of Joe's and some of his clients include Nickelodeon Animation, Dreamworks Feature Animation, Universal Studios, Art Center, Laguna College of Art and Design, and The Academy of Art in San Francisco. Joe is an associate member of the Society of Animal Artists, holds an M.F.A in Illustration from California State University Fullerton. He has also had one man Shows of his oil paintings at Ford Motor Companies Corporate Headquarters, California State University Fullerton and Galerie Daniel Maghen in Paris France. His drawings and paintings hang in private collections in Europe and North America.

SAN DIEGO ZOO
CARACALS
WEATHERLY

of animal form and anatomy, as well as the chops for drawing and painting, one must possess a "passion" for the subject. In other words, to truly capture the nature of animals on canvas or paper, a love for them is required.

When it comes to art, having a passion for your subject is half of the battle. Your work will be better when you have an interest in the subject and a desire to draw and paint it. I've always had a deep interest in the animal kingdom, and animals have been my main choice of artistic subject since I was very young. The tireless study of animals—how they move, how they hunt, how they live—is the path to understanding their nature.

Being able to capture the essence of an animal encompasses many things. The ability to draw and paint animals is the foundational requirement of course, but beyond this ability, one must have a passion for looking at animals, reading about animals, and watching countless documentaries. Knowledge of their habitat and why they have adapted to live where they do plays a large part in the sum total of understanding their essence.

I've said to students that expressing an animal's character is the unteachable part of animal drawing. What this means is that no one can tell you how to draw or paint an animal's nature. It comes as a result of being around animals, studying animals, and loving animals. Expressing the nature of an animal in a drawing or painting goes far beyond technical accuracy, detail, or scientific facts—it's about capturing the essence, the spirit, and the attitude of the animal, and it's worth getting!

Opposite Page:
African Lion
Oil on Panel 6x10
Private Collection

Above:
African Elephant
Watercolor
10x14

Following Page:
The Fox and the Hare
Oil on Panel
12x30

WEATHER

Tigers are among the most popular animals in world, and rightfully so. They are beautiful, powerful creatures. Native to most of Asia, from the hot jungles to the colder regions, tigers are the top predator in their ecosystem. Unfortunately, they are also endangered as a result of poaching and forest destruction. Most of the tigers painted in this book are Siberian, and they are much larger than their jungle cousins. Capturing the beauty of this stunning feline is not easy. With any animal that has a pattern, in this case stripes, one must take great care to make sure that the contours of the pattern follow the underlying form. This is where all of the anatomy knowledge comes in handy. Tigers can range in color from a tawny brown to a brilliant orange. A trip to India to study the environment of the tiger is on my list of things to accomplish, but in the meantime, here are some tigers on abstract backgrounds, painted for beauty's sake.

Opposite Left:
Tigre De Siberie
Oil on Panel 11x14
Private Collection

Opposite Right:
Tiger Head Study
Oil on Panel 5x7
Private Collection

Jungle Stripes
Watercolor
7x10
Private Collection

Tiger Tiger
Oil on Panel 18x24 Private Collection

Asian King
Oil on Panel
5.5x11.5
Private Collection

Amur Tiger
Oil on Panel 9x12 Private Collection

Noble Tiger
Oil on Panel 7x14 Private Collection

Giraffe
Watercolor 10x14

12-1-08
S.A Zoo

10-4-08
SANTA ANA ZOO
GUANACO

11-26-11
LA Zoo

ZOO
CREATURE
.COM
ERLY
ST.
A 92701
ERLY.NET

PAINTING

Much of the art showcased in this book comprises oil painting—my favorite medium for color work. I've said before that, to me, drawing and painting are one and the same. When I am painting with a brush, I am thinking about drawing the entire time. Many artists who draw constantly, both from life and imagination, tend to paint with more athletic brushwork. I am one of those artists. This painterly approach comes from years of putting charcoal or pen to paper, and drawing in a linear and/or tonal fashion. Not only does drawing knowledge lend itself to the development of personalized brushwork, but it also helps as a problem solver for issues such as value, form, and proportions. If one understands tone and how to paint values well, the subsequent color work will be stronger. In many cases, it doesn't matter as much what the color is, as long as the value is correct and the drawing is sound.

I paint using both direct and indirect methods. Direct painting means achieving the end result in one session (alla prima), while indirect painting uses layers to achieve the final result. There are endless ways to start and finish a painting, and artists have a tendency to find what works best for them through trial and error. If I am working on a painting from life, or a small, quick sketch or study, I usually paint it in one session from start to finish. If the painting is an animal scene, group composition, or large painting, I will paint it in a few layers, using over- and under-painting. I rarely glaze between dry layers; rather I prefer to paint alla prima over alla prima. If under-painting is involved, I feel it is always best to finish alla prima to retain a fresh look in the final pass.

For a support, I prefer a wooden panel primed with gesso (and not sanded between gesso layers, as I like some texture to show through in the final painting). I work on canvas from time to time, but I feel that wood (hardboard or birch) lends itself well to "drawing" with a brush, because it doesn't move like canvas does. Also, the firm surface acts as a good support for painting thickly and adding texture. If the painting is large, I will cradle the back of it with a frame for support. In terms of brushes, I mostly use bristle flats, with an occasional sable brush for a smoother passage. Bristles generally hold a lot of paint, and the flats allow me to load more paint on the brush, get a longer stroke, and achieve planar edges.

I work with a set palette most of the time. A set palette involves pre-mixing various piles of color and value, which are then methodically laid out on the palette. Pre-mixing ultimately saves time, because it solves most of the color issues in the beginning. The remaining session can be spent concentrating only on painting and not re-mixing colors. Of course, with this method there will be plenty of adjusting and some re-mixing, but pre-mixing allows for painting more thickly, not running out of color, and bringing unity to the piece when the colors are mixed into one another. I don't use any special colors really, just a warm and cool hue each of yellow, red, and blue, as well as white. I add earth colors and greens to my palette occasionally. I also use a painting medium of one-half odorless mineral spirits and one-half sun-thickened linseed oil.

As with anything in art, there are no rules or absolutes to painting, and I often experiment with different ways start, work, and finish a piece. The process of painting can be difficult, exhausting, and frustrating at times, but when a painting pulls together at the end as something exciting to look at, it's suddenly all worth it. I find that the more preliminary work I do at the drawing stage, the more the paintings seem to "paint themselves."

Ocelot
Oil on Panel 9x12 Private Collection

Next Page: Without Warning
Oil on Panel 12x30 Private Collection

Mountain lions like to prey on deer, but they will also take smaller prey. They hunt with a combination of stealth and power. These cats will stalk until the time is right and pounce on the victim, delivering a fatal bite to the back of the neck. Sometimes they will hide large carcasses in order to feed on them for multiple days. Shown here is a mountain lion standing over its recent kill, a whitetail deer. Although they prefer to hunt at night, sometimes mountain lions will hunt during the day if an opportunity presents itself, as is the case in this painting.

American Lion of the Mountains
Oil on Panel 11.5x24

Desert Lion
Oil on Panel 12x24

Mountain lion, puma, cougar, and catamount are all names given to this powerful predator that inhabits the Americas. Proportionally, these cats have a relatively small head and a very long, full tail. The back legs are long as well, and supply the means of propulsion and leaping power. Mountain lions have spotted fur as juveniles, but lose their spots as adults. Their smooth, reddish-brown coat displays the musculature well and is a joy to paint. Here, I offer depictions in action and repose for the cat of many names.

DRAWING

Many of the drawings in this book were done on location at various zoos. I frequently visit zoos with a sketchbook and draw the animals from life, since this is the best way to capture their nature. The habit of sketching is imperative for artists just learning how to draw, and also for established artists who are honing their skills.

When drawing animals at a zoo, the idea is not to create a perfectly rendered drawing. Instead, try to capture the overall attitude and movement of the animal. Over time, greater knowledge of anatomy and form will inform your decisions, and you will put more information in these zoo sketches, no matter how fast you lay them down. Since animals are usually in motion (even if they are lying down, they will move their heads or adjust their body posture), what you draw becomes a combination of what you see and what you know. A large part actually involves drawing from imagination, even though you are clearly drawing from a live model.

In the studio, while making other kinds of drawings such as a finished piece in a certain medium or a sketch for a painting, one can take more time and use reference as needed. Never be a slave to the reference—use it instead as a supplement to your knowledge. If you understand form, lighting, perspective, construction, and anatomy, you will never really copy a photograph. Instead, you will be able to see things in terms of line and tone, and, hopefully, you will know how to edit by establishing what to put in, and what to leave out.

There are many different ways to draw. I try to experiment with methods, as well as with different types of media. Pen is the most-used medium for me in the field, while it's pencil in the studio. For gesture drawing, I just scribble my way through or make drawings with a very economical line quality. For an average drawing, either from life or in the studio, I will lay in a sketch by getting the action down first. These simple rhythm lines give me the pose, energy, and proportions of the animal in a basic way. This first pass is not set in stone and can be changed throughout the drawing. I rarely use erasers, but rather draw over or draw through the drawing in order to correct or refine it. If I am working in the studio, I will correct by putting a fresh piece of semi-transparent paper over the initial drawing to make the changes and complete it.

After I lay in the action, I start to work on the basic forms. This second pass is a series of very basic shapes and volumes that give me the core masses and structure of the animal. I am constantly thinking about anatomy, and indicating it tastefully where needed. It's important not to overdo the anatomy, otherwise the drawing starts to look more like a diagram rather than a work of art (of course, if I am making an anatomical drawing, it will look very diagramed and structured). Depending on what type of drawing it is, I might leave it alone or I might take it to a more organic, developed stage. This process is known as sketching from the inside out, and it's a sure-fire way to get strong form and sculptural qualities in a drawing.

Draftsmanship, I feel, is very much worth possessing (although not easily earned), because it is the underpinning of so many other forms of great art. Draw well, and it will show in your painting style and ability.

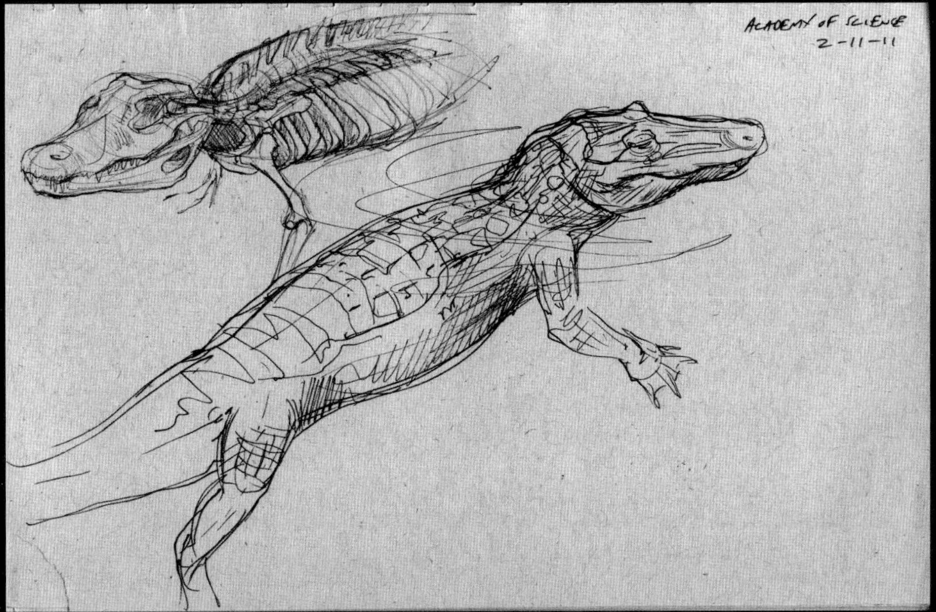
ACADEMY OF SCIENCE
2-11-11

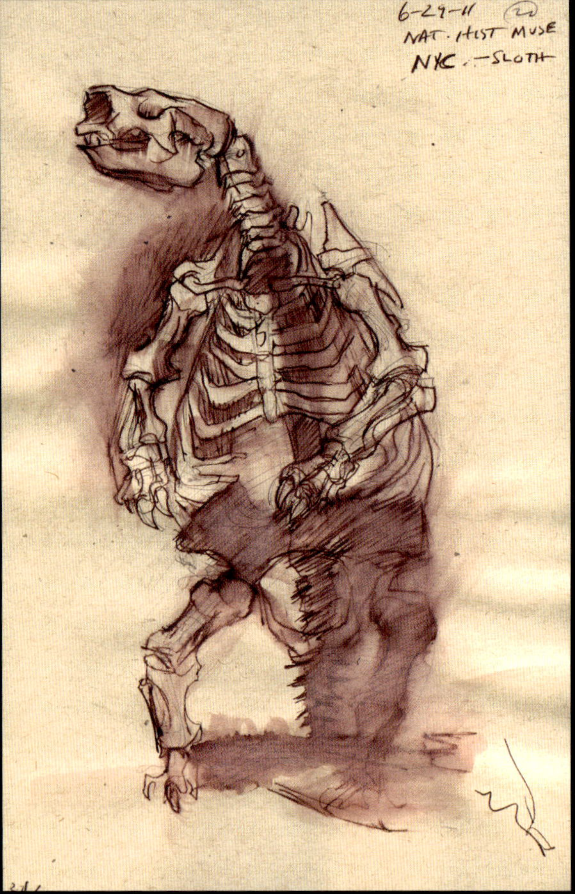

A visit to the natural history museum (in this case, in New York and Los Angeles) always results in me drawing skeletons. Often, my drawings are done from prehistoric models—the information is usually the same as a current-day specimen, only much larger in scale. I try not to approach a skeletal study as if I were drawing a still life, but rather as an armature that will give me the clues I'm looking for in animal structure, proportions, and movement. Shown here is a looser approach done with brush pen, fountain pen with wash, and blue ink on toned paper. I always tell students that the artist who can see the skeleton in their subject already knows a lot of anatomy. The museum is a great place to gain this knowledge.

CANGURA ROJO
8-31-09

8-31-09
DOUBLE WATTLED CASSOWARY

Mexican Wolf
Watercolor 10x14

Cruising Wolf
Watercolor 10x14

Canis Rufus
Watercolor 10x14

Red Wolf
Oil on Panel
14x14

11-14-09
OAKLAND
ZOO

3-4-11
OAK ZOO

8-10-09 L.A. ZOO
GERENUKS

RESEARCH

Research is of primary importance in the success of an animal drawing or painting. Nothing beats going to the source in order to achieve a feeling of authenticity in a painting of an animal and its environment. Obtaining photographs, making field notes and sketches, and—with any luck—a real-life encounter with the animal you are seeking out will all add up in the studio when it comes time for picture making. Some of the paintings in this book are a result of my travels to the jungles of Central and South America, as well as the woodland forests of Montana and California.

Research trips to the animal's habitat allow for first-hand study of the environment. The idea is to build a reference library of landscape elements like trees, rivers, foliage, rocks, and anything else to be included in the picture. For example, to paint a jungle successfully, you need to visit a jungle and see what it looks like under different lighting and weather conditions. Even the sights and sounds of the area will filter back and inform your decisions while making the painting. If time and conditions permit, doing quick landscape sketches and color notes on-site will also prove beneficial upon your return.

Aside from field research, making studies of the animals themselves is imperative. Animals studied on location will give you a sense of how they really look in the wild. This might be quite different from how a well-fed zoo specimen looks. That said, zoos offer a convenient way to study animals, as do natural history museums. In the pages that follow, you will find paintings of Latin America's largest predator, the jaguar. To research this great cat—almost impossible to spot in the wild because of their reclusive nature—I spent many hours at zoos drawing them, painting them, and watching the way they move. When I am looking at them, I study how the skeleton moves, how the light hits their coat, how their rosettes are arranged, which muscles are most prominent when they are walking, and so on. At home, I research the muscles and bones of big cats and fit these drawings into the proportions of the jaguar. I also watch film footage and read about their lifestyle and habitat. This part of research is another way to understand an animal's character. I try to use this approach for most of the animals that are included in my paintings.

Research for the actual painting will include idea sketches and thumbnails. If the painting is a go, I will proceed to work out the composition of the piece and draw it out. For color research, I usually do at least one, if not more, small color sketches. For a group composition or a painting with an elaborate landscape (such as Mayan ruins or temples), I make small, individual painting studies of the particular species or landscape element. The process of painting the final piece is very spontaneous, but I believe that the more research I do ahead of time, the better the chances for creating a strong, successful end result. Research is an enjoyable process for the animal artist, and well worth the effort.

perú
MININTER - DIGEMIN
01 JUN. 2011
SALIDA

Most of the paintings I make start with small, scribbled thumbnails. During this part of the process, I am not overly concerned with drawing perfection, but rather getting down a good idea that is worth developing further. Shown here are a series of small thumbnails that evolved into tonal drawings made with markers. I moved the cat and the ruins around several times before deciding on a final composition. Once that decision was made, I proceeded with the color sketch, and then the final painting.

I wanted to make a painting about a jaguar walking the stone steps of a great Mayan temple. Although not a common occurrence, jaguars have been known to enter or walk the perimeters of these jungle pyramids, and the pre-Columbian Mayan civilization had multiple jaguar gods. A trip to northern Guatemala to explore and photograph the National Park of Tikal provided the reference for the jungle temple and surrounding foliage. This painting is a combination of imagination, reference, and knowledge. Care was taken to achieve a pleasing composition that has balance, unified lighting and proper perspective.

Exploring Tikal
Oil on Panel 36x48

Lords of the Jungle
Oil on Panel 36x48 Private Collection

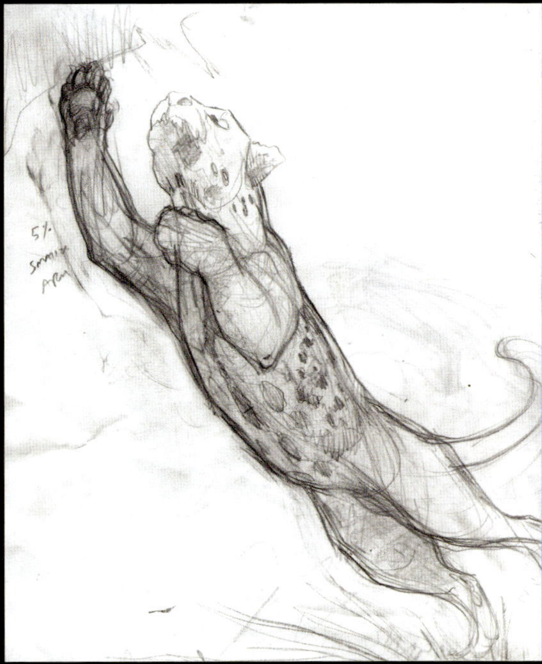

Narrow Escape
Oil on Panel 24x30
Private Collection

In the painting "Ambush," excitement and fast movement run the show. Action-oriented paintings require one to marshal up knowledge of active anatomy, rhythm, dynamic composition, and imaginative powers in order to be successful. If a painting depicting animal motion comes off as being stiff, the movement itself looks frozen. Techniques that enhance the illusion of movement in a painting include: pushing the pose, using paint strokes to lead the eye in and out of the composition, and using soft edges or a blurred effect in key areas to indicate speed. Shown here are the various stages of preliminary work, including idea drawings, a tonal comp, and a color sketch. Also included are pictures of jungle landscapes from the Amazon River in Peru, which provided the backdrop for this dramatic scene. The final piece depicts a jaguar, mid-air from a powerful leap, ambushing a pair of snowy egrets that were peacefully wading in the jungle pool. Whether or not one or both of the birds make it to safety is for the viewer to decide...

As a subject for painting, the predator/prey aspects of animal behavior and, more specifically, the chase have always fascinated me. The heavy drama involved in these scenes, as expressed through the dynamics of action and rhythm, is exciting to paint.

Ambush
Oil on Panel 18x30
Private Collection

Above are the first thoughts on paper of how to show the jaguar and caiman in a savage confrontation. Below are some painting studies of the yacare caiman, and reference pictures taken on location in Brazil. Also shown is the color sketch. The spectacled caiman was chosen for the final piece because I thought it harmonized better with the warm river colors.

"Forces of Land and River" comes from Mayan legends of earth and water, the jaguar, and the crocodile. At any rate, a jaguar trying to tackle a half-submerged caiman is going to be a "forceful" event. The scene was conjured up from my imagination, but based in reality. In the tropical waters of their jungle homeland, jaguars quite often hunt and kill big reptiles, such as caiman and crocodiles. The circular composition of the piece keeps the eye moving around the action from top to bottom, while the opening between the upper-corner trees provides a means for the eye to escape. An impressionistic (rather than rendered) approach is used to handle the scales on the caiman to avoid an unnecessarily busy area. Deft strokes with the knife and brush create the illusion of the tough, crocodilian skin.

Forces of Land and River
Oil on Panel 36x48 Private Collection

When we think about where jaguars live, usually Central and South America and Mexico are mentioned. However, the jaguar also once inhabited parts of the American Southwest, including California, Arizona, and New Mexico. These cats roamed from Mexico into the borderlands of the United States until they were all but removed by poaching and habitat destruction. Recently, jaguars have been seen on American soil crossing into the deserts of Arizona from Sonora, Mexico. This painting is an imagined scene of just that. “Border Crossing” shows a male jaguar wandering over the United States/Mexico border into the saguaro cactus–filled deserts of Arizona.

Jungle Queen
Watercolor
10x14

Jungle Cat
Watercolor
10x14

Panthera Onca
Oil on Panel
12x24
Private Collection

Opposite:
Border Crossing
Oil on Panel
16x24

Survival
Oil on Panel
13.5x24
Private Collection

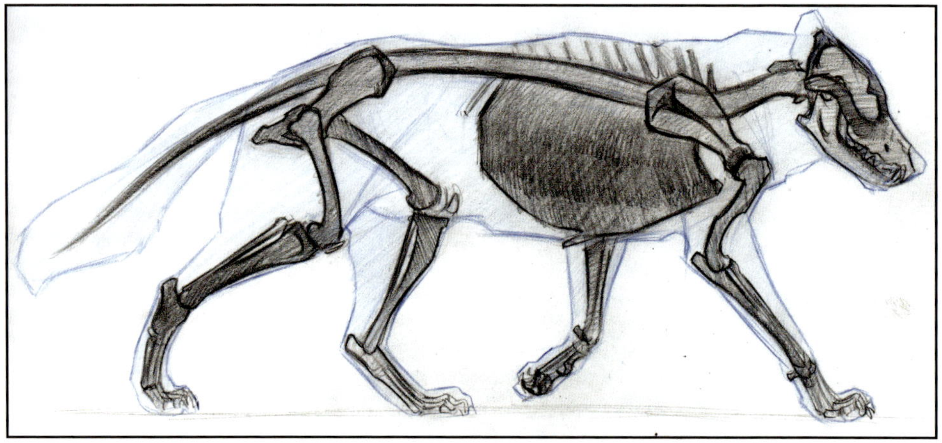

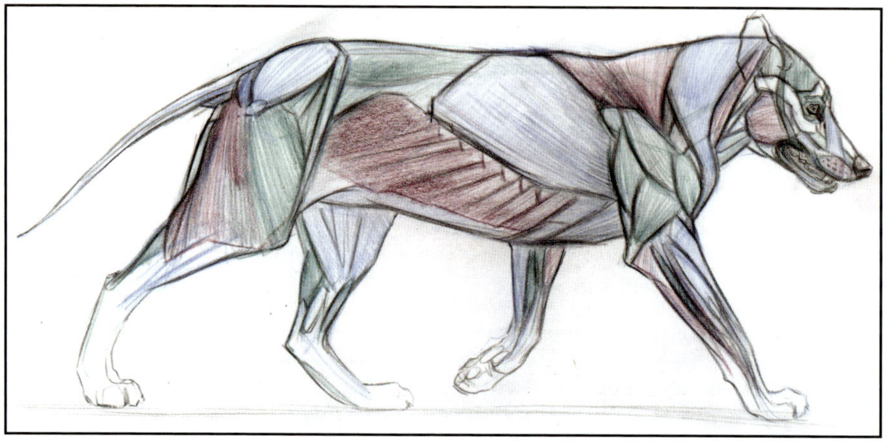

The research shown here comprises a complete study of a Mexican Red Wolf, including a skeletal and muscular study, an exterior pencil drawing, a color sketch, and the final painting. Knowledge of the animal's anatomy really pays off in the final result. The skeletal study establishes the framework and proportions of the wolf. The musculature hangs on the framework and supplies clues to the many volumes and surface forms. I don't always do a small color sketch for a portrait. When I do, however, I find it makes the job of mixing paint for the final piece much easier, since many of the colors have been worked out in a general way.

Mexican Red Wolf
Oil on Panel
12x24

Opposite:
Old Friends
Oil on Panel
16x20

Testing the Waters
Oil on Panel 14x20
Private Collection

Roaming
Oil on Panel 20x32
Private Collection

Asian Elephant
Watercolor 9x12

Mountain Cat
Watercolor 10x14

Cheetah
Watercolor 7x10

Snow Leopard
Watercolor 9x12

Siberian Tiger Watercolor 10x14

MTN
GOATS
LA ZOO
2-5-10

Previous Page:
Springtime
Oil on Panel 12x30 Private Collection

WEATHERLY

WEATHERLY

Coyote
Oil on Panel 16x20
Private Collection

Wolf Study
Oil on Panel 6x9
Private Collection

Snow Wolf
Oil on Panel 9x12
Private Collection

Next Pages: Grey Wolf Oil on Panel 16x18 Private Collection, and White Wolf Oil on Panel 16x18

Indian Elephant
Oil on Canvas
9x12

Lioness
Watercolor 6x9

Resting Lion
Watercolor
6x9

Lion
Watercolor 7x10
Private Collection

The noble lion has been known throughout history for its strength and courage. This king of beasts inhabits the savannas of sub-Saharan Africa. Female lions, which perform most of the hunting, do not sport the large mane that circles the head and neck of the male. When drawing a male lion, this neck area must be visualized beyond all of that fur, otherwise the proportions will go askew. On the left is a series of quick watercolor paintings of male and female lions, done both on location and in the studio.

WEATHERLY

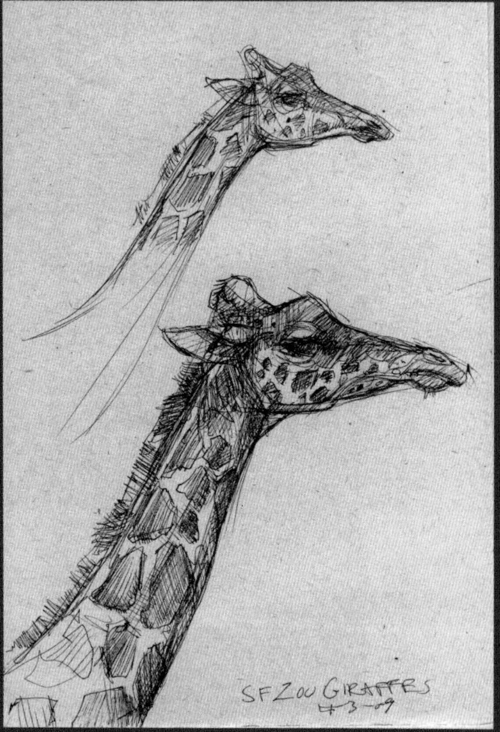
SF ZOO GIRAFFES
4-3-09

Black Rhino
Watercolor
7x10

Hook-Lipped Rhino
Oil on Canvas 9x12

Black Bear
Oil on Panel
14x14

Brown Bear
Watercolor
7x10

Bears are a zoo favorite of mine. Their playful antics and robust, round forms make them a delight to draw. Shown here are the black bear—less massive than the grizzly, and with a smaller shoulder hump—and also the polar bear. Bears have shaggy fur that conceals much of their underlying bone structure, but if you look hard enough, you will start to notice all of the "landmarks" coming to the surface. A bear drawing is also a good lesson in the design principle of straights against curves.

Finding a niche in art or a specialty within your chosen field is a great way to develop a name for yourself. I chose long ago to focus on animals as the main subject matter for my artwork, and I have stayed with it because the inner passion I have for drawing and painting them still burns brightly—in fact it burns brighter than ever! Like anything done over and over again, year after year, things can seem monotonous at times, but persevering and embracing it as a lifestyle is the true road to mastery of a subject. The love of the subject matter keeps things exciting. The nice thing about animals is the almost endless variety—there is no shortage of subject matter within the animal kingdom, and this keeps things fresh.

This book is the continued journey of my animal art. I hope you have enjoyed looking at the images collected here. I have most certainly enjoyed creating each one, and I look forward to filling up many future sketchbooks and canvases with the subject's inherent constitution of character otherwise known as "animal nature."

Also by Weatherly Studio

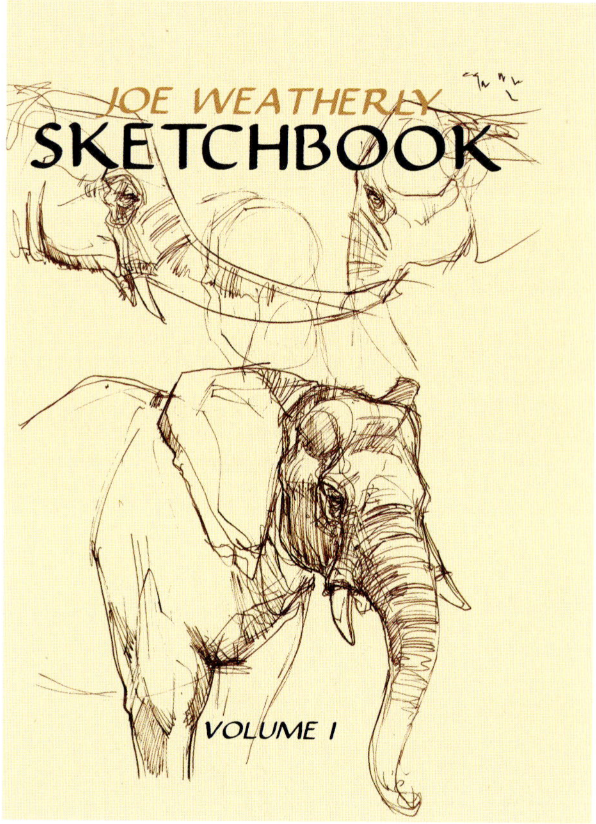

For more information visit
www.joeweatherly.com